Though the words
in this book *may* modify
your view of yourself
and the world,

they do not constitute
a philosophy per se,

nor do they in any way
constitute a religion
or a belief system,
a prediction
or a prophecy,

they are, basically,
a simple, objective,
statement of fact.

The One Truth
(amended edition)

ISBN-13: 978-0-557-99329-1

fabianmelmelgar@gmail.com

# THE ONE TRUTH

Fabian Mel Mel Gar

There are
many theories
about man,

and many
beliefs about
the gods;

there are many
tenets about the
heavens,

and many
truths about the
universe;

but there is
only *one* truth
that *governs* man,

*and* the gods,

*and* the
heavens,

*and* the
universe;

**only *one*,**

and that truth
stands on two
eternal principles
that determined
our past, our present
and our future:

the first is
that *all* motion
moves *forward*
in time,

the other is
that *all* things
that occur,

whether
predictable, or
not predictable,

are *caused*;

and these two
principles are
embodied in this
*one simple truth*:

*Everything*
**as it is**
**at this**
**moment**

was
caused by
*everything*
**as it was**
**a moment**
**ago.**

*That* may
seem an
innocuous
truth,

but consider
its implications
from a
l o n g e r
p e r s p e c t i v e :

*Everything*
**as it was
a moment
ago**

was caused
by *everything*
**as it was
the moment
before that,**

and
*everything*
**as it was**
**at *that***
**moment**

was caused
by *everything*
**as it was**
**the moment**
**before *that*,**

and
*everything*
**as it was**
**at *that***
**moment**

was caused
by *everything*
**as it was**
**the moment**
**before *that*,**

and
*everything*
**as it was**
**at *that***
**moment**

was caused
by *everything*
**as it was**
**the moment**
**before *that*,**

and back
and back
and back
and back
and back
and back
and back
and back
and back
and back
and back
and back
and back
and back
and back
and back
and back

to the

**beginning of**

**everything.**

And though, logically,

for there to have been

a <u>literal</u> *beginning*

*of everything,*

there would first had

to have been a time

when there was

*nothing*

—no matter or energy,

known or unknown,

no planets, no suns,

no galaxies, no

universes, no gods,

*no anything*,

and then, from that

absolute and

total nothing,

*suddenly everything*—


we can just arbitrarily,

for the purpose of this

narrative, choose

any point in history

and call it

*the beginning*;


and it doesn't matter

whether we choose,

five thousand,

or ten million,

or seven billion, or

eight centillion, one

trillion, ten million and

fifty seven years ago...

*Everything*
**as it is at
this moment,**
and *everything*
**as it will be at
any moment
in the future,**

was, and will be,
*caused* by
*everything*
**as it was at the
<u>BEGINNING</u>.**

*It cannot be otherwise.*

# And *everything* means *everything*

–past, present and future–

including,
but not limited to:

the exact moment of your birth,

the precise moment of your death,

all the tears you've shed,

all the smiles you've smiled,

every twitch you've twitched,

every itch you've scratched,

every fall of a leaf,

every burst of a bud,

every random event,

every event accidental,

every story in The Bible,

every word in The Koran,

every road to Brahman,

every path to Nirvana,

every statement by Confucius,

every explication by Darwin,

every word ever written,

every book ever hidden,

every obstacle ever climbed,

every rhyme ever rhymed,

—everything and everything
and everything—

*everything* means *everything*,
**nothing is excluded.**

And, it was all *caused*
by everything that preceded it,
all the way back to the
beginning of it all.
It cannot be otherwise.

*How can it be
otherwise?*

*(When I say* everything *as it is at this moment was* caused *by* everything *as it was at the beginning of it all, I don't mean to say that it was planned or designed by some intelligent being, or beings, as many people believe —this is not a book founded on beliefs, it's a book based on facts and logic that constitute a truth— I'm simply saying that* everything *as it is today was* causally determined *by everything as it was at that beginning; and that includes anything attributable to any god that may exist, because if there is a god, what that god does at any given moment is caused by everything that preceded it. That god's decision to create us, for example, would have been caused by the state of his mind the moment before he did so —he couldn't have created us and then decide to do so. Of course, he could have created us by accident, but that too would have been caused by everything leading to the accident.)*

# And *caused* means *caused* —nothing *just happens.*

———•———

Place a small BB on a flat board so that it is perfectly still relative to the board. Blow on the BB and the BB will move; nudge it with something and it will move; place a magnet in it's vicinity and it will move; tilt or shake the board and the BB will move; it will move because one of those actions —or some natural process such as oxidation— will have *caused* it to move; none of its movement could have just happened *of its own volition* —if you were to wait for it to move on its own you would have to wait forever. Of course, if there is a god, one would assume that he/she could make the BB move; but, if so, that god would be the *cause.*

*The one truth* may be easier to accept for *large, static* objects than for *tiny, erratic* ones.

On a clear night, look up at the moon as it circles our planet on its delineated orbit. If you were told that its exact position, at that precise moment, was *caused* by everything as it was at the beginning of the world, you could probably accept that possibility. But if you were to look up on a sunny day and see a little yellow butterfly fluttering by erratically, and were told that *that* specific butterfly and its exact position, at that precise moment, was *caused* by everything as it was at the beginning of the world, you would be loath to acknowledge that. Yet it's true, and there is no way it can be *not true*.

# If you say a prayer

for the life of a loved one who is ill, and that person survives their illness, it doesn't matter whether you think that your god saved your loved one, or you think the doctors, nurses, medicines or technicians saved your loved one, or that your loved one's immune system did it, the cure was caused by everything as it was at the beginning of everything. But if you choose to say, that in response to your prayers your god changes what it was determined he would do, it is simply not so. If there is a god who answers a specific prayer, it was determined, by everything that preceded it, that that specific prayer would be made and that that god would respond to it according to who he/she is at the moment that that specific prayer is made and the specific nature of that prayer.

*It cannot be otherwise.*

Everything as it is at a particular moment was caused by everything as it was the moment before —*and nothing can change that truth, not even a god.*

# Your choice of a mate

was irrevocably determined before humans existed. You had nothing to say about it, you couldn't have chosen differently. A confluence of events, going way back to the beginning of it all, brought you together, and who each of you were, at the moment you met, sealed the union.

**Consider this:** there are more than 100 trillion living cells in each of our bodies, and they are all in continuous motion —internally and/or externally.

And we and our bodies move on a planet that contains approximately 7 billion of us. So that's 7 billion living bodies times 100 trillion cells, or a total of seven hundred sextillion, or 700,000,000,000,000,000,000, 000, living, human cells on our planet that are in motion at every given moment.

And there are, on earth, conservatively, as best as I can determine, 60,000,000,000,000, 000,000,000 individual, living, non-human mammals, reptiles, fish, birds, insects, plants, bushes and trees, and slime, molds and other fungi; and let's say, very conservatively, that there are, on average, 100 million cells in each of these individual living things. So that adds up to a total of 6 nonillion (that's 6 with 30 zeroes after it) non-human living cells. Add to *that* the 700 sextillion human living cells for a total of 6,000,000,700,000,000, 000,000,000,000,000 living cells in motion on our planet every moment. And add, to all those living cells, the huge quantities of animate bacteria and viruses, and the even larger quantities of inanimate matter that constitute our planet, and the fact that each of those living cells, and each of those bits of inanimate matter, is made up of trillions of molecules, which are made up of atoms, which in turn, are made up of particles —all in constant motion— and we have an incredible, unimaginable amount of moving particles on just our tiny planet.

Add *to that* all the particles in all the other planets, moons, meteors, comets, rocks, and pebbles that whirl around our sun, and our sun itself; and multiply *that total* by the 100 billion suns, and their satellites, in our galaxy; and multiply *that total* by the 200 billion or more galaxies in our known universe, and add to that the hypothetical dark matter, which, it is said, lies between galaxies and galaxy clusters, and represents the bulk of the total matter of our universe. And multiply *that total* by a boundless number of other universes that surely must exist; and we arrive at numbers we cannot possibly conceive of, dizzying numbers we have no name for. And that unnameable number, which doesn't include any gods that may exist because we can't know how many moving parts there might be in a god, would only be enumerating everything as it is at just *one moment in time,* one moment of a staggering number of moments that have already existed —staggering even if we only go as far back as the big bang, some 14 billion years ago.

Now imagine those enormous, unnameable quantities of

particles of matter flowing
through that staggering number
of moments of time; attracting,
repelling, crashing, merging with
each other, bumping, nicking,
kissing, bouncing, ricocheting
and splitting till arriving at their
*exact position at this exact
moment;* all caused by their
exact position the moment
before, and their exact position
the moment before that,
and before that, and back
to their exact position at
*the beginning of everything*.

Seems incredible, but it's true;
not one thing as it is at this
moment can be different than it is.

### *Not one.*
No exceptions,
no maybes.

# But now, imagine,

if you can, the 'you' that inhabits your brain. What is this mind or soul that is you? How does it work? Is it the result of evolution? Was it put there by a god? Whatever you think initially created the 'you' that you are, the 'you' that you are at this moment was caused by *everything* that preceded this moment. Many people around the world believe that a god created your soul, or the 'you'

that you are, and put it in
your body, and that this
'you' is an ineffable thing
with no material basis
—though, if I'm not
mistaken, most accept
that it resides in the brain.
Others, those with a more
scientific understanding of
our mind (or soul) see it
as a *material* thing in the
same sense that the bits
of information in a com-
puter have been put there
by *physically* altering
the ferromagnetic surface
of the disk in its hard-

drive. That is not to say
that they claim the mind
works in the same way,
it is just to say that the
*mind* is a *material* thing
just as bits of information
in a computer are *material, retrievable* things.
But regardless of how
you view your mind or
soul, everything it does at
a given moment is caused
by the state it was in
the moment before.
It cannot be otherwise.
You could not exist
if it wasn't so.

*If* you've read this far, it's because it was caused by the words that preceded it. Each word that you read followed the preceding word that you read, which followed the word that you read that preceded it. Each word you had read caused you to read the next word that you read and caused you to have thoughts about what you were reading; and the specific thoughts you were having were determined by who you were at the moment you were reading them.

Whether, or not, *you* are able to understand any of that, was determined long before the beginning of thought.

"So what?"
you may ask.
"Enough already,
I understand
that everything
as it is
was caused by
everything that
preceded it.

But who cares?

What does it
mean to *me*?"

It means that you did not create the 'you' that you are at this moment, it means you were created by all the things as they were in all the moments that preceded this moment all the way back to the beginning of everything;

it means that you were created by a god, or/and, your ancestors, your parents, your biology, your family, society, and all you've experienced; and the exact person you are —at this moment— is **the *only* person you can be.**

And that means:

# you
# don't
# have
# *free*
# will

—that is, you are not capable of making a choice different than the one you make. **It means you have will, and it's yours, but it is not *free* to choose other than what it chooses.** It may feel to you, as you live your life, as if you *are* making choices; it may feel to you as if you are choosing between two or more options; —and you are, but...

the choice
you make
at any
given
moment
is the
only one
you *can*
make.

We humans, masters of our world, don't want to know that we can't make *free* choices, that the next choice we make will not be of *our own* making, that it will be caused by a long chain of past events leading to the moment we make the choice. We feel that to admit *that* —that we have no *free* will— means we are admitting that we are nothing more than mindless automatons; that we'll obey who

we're told to obey; that we'll *believe* everything we're told; that we'll *do* anything we're told; that we'll *love* who we're told and *hate* who we're told; that we'll believe the most childish fantasies; that if we have no *free* will, no will of our own making, we will be nothing more than placid, malleable sheep induced to follow any Pied Piper who promises to lead us to green grass and cool water.

And we
would be right.

All those things
*can* happen,
and *have*
happened,

*not because*
**there is**
**no _free_ will,**

*but because*
**we _think_**
**there is.**

Most of us believe we are
*free* to think whatever is
thinkable, and do what-
ever is doable; and we
believe it because we
think there is a separate,
inviolate, autonomous 'me'
that resides in us, a
distinct 'me' that's there
from the moment we're
conceived, a 'me' that is
virginal and pure and not
influenced by its creator
—be its creator a gene
or a god or both— an
individual 'me' that's able,
from the moment we're
born (or perhaps before),
to use its supposed *free*
will to make choices —to
discern whether incoming

information is factual
truth or spurious fiction
or something between.
But, there are others of us
who believe that except
for our genes —which
determine our physique,
the construct of our brain,
and, probably, some, or
many, of the rudiments of
our mind— there is no
*autonomous* 'me' when
we are conceived, or born,
only the gradual emer-
gence of an awareness of
ourselves as we develop,
first in our mothers womb
and, later, outside of it.
And, lastly, there are
those of us who believe
that we are born as

naked in mind as we are
naked in body, with no
sense of anything till we
learn from the world
we're born into.

But it doesn't matter
whether you think of
your 'me' as your 'soul',
or just your 'me'; or
whether you believe you
are born with a blank
brain; or with a mind
already having begun to
process and use incoming
information; or with
something between; we
need someone with expe-
rience of the world to be
our guide, to inject our
mind with contemporary

knowledge that will teach us how to survive —each according to our innate abilities— in the specific world we are born into. And that someone is our parents and our siblings and anyone close who is active in our lives; and it is *they* who teach us about the world, and they teach us according to *their* beliefs and under- standings; and since there is no one in our lives as powerful as *they,* who can contradict what *they* are teaching us, *their* ideas —remodeled by our needs and abilities— become imprinted indeli-

bly in our minds. Then, as we grow and we come in contact with the wider world, our imprinting is strengthened, or modi-fied, by our teachers and our playmates, and, later, also by books and maga-zines and newspapers, music, movies, television and the internet. And, after all that —when we are finally adults— the 'we' that we are is the product of our biology and all that imprinting; and, unless we have been taught to keep our mind open and critically questioning, that imprint-ing becomes set, like

concrete, in our brain. But regardless of whether our brains are set in concrete or our minds are still pliable, what we are at a given moment is what we are and we cannot be otherwise, and what we do at that moment cannot be other than what we do. But when we think there is an inviolate 'me' that is not *ruled* by what we've inherited, and —more apropos here— what we've been imprinted with, we don't question what we are doing, we don't question the imprinting we have

received as children and the indoctrination we've continued to receive throughout our lives; and *that* may cause us to do things that are not in our best interest, or in the best interest of those we care about, or in the best interest of the society we live in, or that of the world at large. And that is the danger. The danger is not the fact that *we don't have free will,* but that ***we <u>think</u> we do***; and that because we think we do, we don't develop the tools of critical thinking and questioning that would allow us to make

*informed and examined* choices instead of the *sheep-like* choices dogmatic indoctrination leads us to make.

And the biggest danger of all is that because we believe we have *free* will, we pass that belief onto our children and we *don't* teach them to bring an open but critically questioning mind to the consideration of new ideas. Unfortunately, that leaves them helplessly stuck, for a lifetime, in a bog of antiquated, narrow beliefs, or captive to a whirl of new fads foisted

on their unquestioning
minds by promoters of
fashionable products or
philosophies.

And for those of us who
think we have *free* will
because a god gave us
our soul, and, with it,
*free* will, it must be that
when this god speaks of
*free* will, he simply
means that he gave us
the ability to make choices; because, if this god
knows *all*, he knows that
we *can* make choices but
that the *only* choices we
*can* make are the ones
we make; and that is not
the *free* will that most of

us would like to believe we have. The 'we' that we are at  the moment we make the choice —and the event that requires the choice to be made— determines the choice, and that choice is the *only* one we can make.

*Our brain, with its mind, or soul, is the only tool we have that we can use to make choices, and that tool can only be as it is, it can only be as it was crafted for us.*

(That does not mean, however, that we are all the same —as if we came off an assembly line; it does not mean that we don't have a will that's our own. Quite the contrary, we are all unique: we are all —including identical twins— born biologically different, and —except, possibly, for twins— at different points in our family's and the world's history; we may live in different areas of the world, and have different belief systems, and be in different economic and political environments, and have different experiences; all of which possible combinations, if totaled up, would add up into billions of trillions of different possibilities, making it quite impossible for there to be two of us humans exactly alike, with identical wills.)

# It is not possible for us to change *on our own* the 'we' that we are

—even if our brains are not set like concrete, and even if we *want to* change the 'we' that we are. Only *new knowledge* and *new experiences* can change who we are. Sometimes a new bit of information that we are exposed to can fill in the gap between two seemingly unconnected pieces of knowledge that we already possess, creating something new in our minds —an important new vision of the world for

instance. And sometimes an experience can dramatically, and forever, change us and our vision of the world regardless of the state of our mind, our will, or our wishes.

Take the example of a strong, young man who thinks himself heroic. One day he is in a local grocery when a hooded man suddenly points a murderous looking machine gun at him and the others in the store and tells them he will kill anyone who moves. As the desperate hold-up man tells a clerk to get him the money from the cash register, our

hero decides —that is, he chooses, because of who he is at that moment— that at the first opportunity he will jump the hold-up man and wrest away his gun. Suddenly, at the sound of a police siren, the hold-up man turns to look out the window, and our hero is about to jump him when another young man, also envisaging himself a hero, dives at the hold-up man. But before this *new* hero can get his hands on him, the hold-up man turns and fires the machine gun at him, effectively blowing his head to pieces and killing a

neighboring person in the process. Suddenly, on witnessing this, our original hero becomes a different person than he was. Some might say he has become a coward; more likely we should say he has been profoundly changed, and not by use of his will but because the experience has made him a different person, a pragmatic realist; and when the killer escapes out the back and is never caught, our hero also becomes a cynic —one less able to have faith in the ability of society to protect him.

That *the one truth* applies to the 'so-called' immaterial world, as much as it does to the material one, may be hard for many to accept.

Those readers who may have the most difficulty grasping the truth of this absolute determinism are those who believe in the existence of ineffable, transcendental, ethereal things like goblins and ghosts and

spirits and souls, and rein-
carnation, and gods and
angels, and the efficacy of
prayer, and heaven and hell,
and in the ability of some
humans —those with extra-
sensory perception— to see
into the future and the past
and to contact the souls of
the dead. Those readers feel
certain that those things
have no material aspect, that
they are some form of supernatural magic, and they can't
imagine how *the one truth*
could apply to them. And, as
to their god, they will say,
"Our god is all powerful, he
can do anything, *he* has *free*
will." That a god is all powerful in every other sense

may or may not be true, but *the one truth* is an unequiv-ocal truth that applies to gods as well as to ourselves, and to souls and ghosts and spirits and angels and all other things that appear to have no material aspect. It can not be otherwise. Neither gods, nor souls nor any other *immaterial* thing can do something *before* they think about it —they can't scratch their butt and *then* think about doing so. And the truth is that not even a god can alter *that* any more than *we* can. A god cannot have a will that is *not* predicated on forward sequential thought,

because without a forward step-by-step progression of thought and action, this god would not be able to accomplish anything —he could create no us, no earth, no universe, no anything. Many people think of gods and the soul as being supernatural and immaterial; but, if the heavens and gods and the soul do exist, they are not *nothing*, they are *something*. The fact, as claimed, that these things are not analogous to the material world does not mean they are not subject to *the one truth*. And if there is a hell, it also is not *nothing*, it is *something*; and it cannot be other than what

it is because hell, too, must observe *the one truth.* Neither in heaven, nor in hell, nor on earth, can you pick your nose, or take a step, before you consciously, *or unconsciously,* think about doing so. If *the one truth* weren't a fact, the heavens would be incapable of being, and gods' minds would be incapable of thought, and the same would be true of our souls or our 'me'.

That brings us to another thought: people will say, in contradiction to the fact that the brain works in a sequen-tial, progressive thought pat-tern —our minds arriving at

a conclusion about some-thing *after* thinking about it— that they often sponta-neously come up with a solution to a problem, or spontaneously react to some sudden physical danger, without being aware of any thought process preceding it. While it may be true that there is no thought process in the seemingly sponta-neous reaction to sudden physical danger, it's because consciously *thinking* about the danger would take too long and could possibly be fatal. I am not a biologist but I would assume there is some automatic instinctive response that is an inherent

part of us, that is crucial to our survival, that causes our reaction —as would be the multi-tasking ability of our brains that allows us to consider one problem while unconsciously mulling over, and coming up with the solution to, another one. But regardless of how you think the brain and its functions work, the process always flows progressively forward, moment after moment. There is no way it can be otherwise, brain processes cannot *progress* backward in time.

Nothing can.

Here's a fact that may
help explain one aspect
of the one truth:

# Time is a measure of motion.

Everywhere in our universe
there is motion. From the
tiniest particle to the largest
galaxy everything is in
motion internally and externally —it is all just part of
the counterpoising forces
of the universe at work—
*and 'time' is one of the
measures of the progressive
motion of those things.*

If there were absolutely no

motion in the universe —something scientists say is physically impossible— the universe would be nothing more than a collection of static particles, everything would be in frozen equilibrium, there would be no *time* —as a measure— since there would be no motion to measure, and, in effect, if not in reality, no universe since there would be nothing in motion in it. So there *is* a universe because there *is* motion, and motion is always forward, always progressive —nothing can progress to moment B before first being at moment A. Some may want to argue that things *can* move backward;

and that's true in a parochial sense: they *can* move backward as when an automobile is backing up; but they are still moving forward in their progression of motion —the time indicated on the automobile's clock will continue to progress forward while the automobile moves backward. Whether an automobile is moving from east to west nose first or butt first, it is still moving forward in time from where it was.

There are imaginative, popular science writers who speak of *physically* going back in time to the ancient past or forward into the distant future. They say that it's possible. They even say that time can

flow backward —effects happening before their cause. While these and other even more exotic speculations are apparently not disallowed by some scientific interpretations of *Einstein's theories of relativity,* there is no reason to believe that any of those things have ever occurred, or ever will occur, or that man will ever be able to cause them to occur. For now, none of that matters, everything as it is at a given moment was caused by everything as it was the moment *before* —or maybe it will be the moment *after,* if, sometime in the far distant future, time should begin to flow backward.

# At this point,

you may be saying to your-self that a lot of this about the one truth may be true, but that, nevertheless, you feel that *you* have *free* will and that you can change your mind anytime you want to and do things differently than you might do otherwise.

To prove it, one Saturday morning you get up earlier than you normally would —deliberately changing your routine. You go to the bath-room and sit on the toilet longer than you typically would; you take a longer, hotter shower than normal;

you get dressed in exercise clothes you haven't worn in years; and you go for a morning walk —something you've never done— in a wooded area that you've never been to; you follow a well-worn path that seems to circle around through the woods and back to where you started; but after you've been walking a while you come to another path —a path, not so well worn, that seems to turn off deeper into the woods— and in order to be different, in order to prove to yourself that you have *free* will, you take this less traveled path; and sometime later you suddenly, unexpectedly, come upon a big momma bear caring for her two cubs. She turns your way and instantly determines that you are a threat to her cubs and she is on you before you can react, knocking you down, slashing your throat, going for the jugular. Then she leaves you there bleeding to death. And so, you would say, if you were capable of saying anything, "See, I changed what it was determined I would do, and I changed the moment it was determined I would die." But the fact is that you wouldn't have changed anything. The fact that, hypothetically, you had been in a bookstore and

were intrigued by this book,
and had begun to read it, is
because of who you might
have been as a person at that
moment in your life: the kind
of person that led you to be
in the bookstore in the first
place, the kind of person who
wanted to read this book,
and the kind of open-minded
person who liked to disprove
other people's assumptions;
and because of that you got
up early that morning and
did all your morning rituals
differently and went for a
walk in a place you had never
walked before and took an
alternate, lightly used path
deeper into the unknown and
found yourself face to face

with death. That's the person you were when you picked up this book, and you could not have done anything differently; everything as it was at the moment of your death was caused by everything as it was the moment before, including every thought and emotion in your brain and the existence of this book and of that bear. And, whether this were true, or just the story that it is, the moment of your real or fictional death would have been determined a long, long time ago, long before you entered that book store —long before man was man or bear was bear.

*This book* may have, by now, convinced you that you can't change who you are or the things you do, and so you may say to yourself,

*"If everything that has happened and will happen in my life is causally determined and unchangeable, I might as well stop striving to accomplish <u>anything</u> and just sit on my hands and do nothing."* Well, if that is who you are and who you want to continue to be, *that* is what it was determined you would do; but if, instead, you go on striving to do something productive with your life because that

is what you wanted to do, then *that* is what it was determined you would do, and the life you lead will be a result of that. In either case, you could only act according to what you are and what the world is.

The above also applies to any god that may exist. *That* god might one day choose —in order to be different, in order to demonstrate that he has free will— to terminate our universe and create an entirely new one. If he does so, it's because of who he is at the moment he makes that

decision, and he would be the cause of all the effects of that. He could, alternately, because of who he ostensibly is, completely eliminate our universe as if it had never existed, as if it never occupied a slice of time or a space in his own memory. But if he did that, he would then be at the same state he had been at when he first created our universe and he would do everything the same as he had done it before, creating our universe exactly as he had done it before, and causing you to be reading these exact words just as you did before.

This may not be apropos
to the subject, but

# have you ever
# stopped to think

that if *just one* of the hundreds,
or hundreds of thousands,
of your direct ancestors:

your mother and father

and *their* mothers and fathers

and *their* mothers and fathers

and *their* mothers and fathers

and *their* mothers and fathers

and *their* mothers and fathers

and *their* mothers and fathers

and *their* mothers and fathers

and *their* mothers and fathers

and *their* mothers and fathers

and back and back
as far back as you can go,
had not reproduced —if just
*one* of those couples had not
met and been drawn to each
other and joined together
sexually— you would not
be fortunate enough to be
here reading this today.
But *here you be*, and,
just by your presence,
creating a world different
than it would have
otherwise been, while the
vast, vast majority of the
innumerable people that
*could* have *possibly* been,
*never were and
never will be.*

# But here is something that *is* apropos:

When a man ejaculates his semen into the vagina of a woman, it contains as many as a quarter of a billion sperm which immediately proceed to try to find their way to the egg in the woman's fallopian tubes which may, or may not, be awaiting the winner of this genetic olympics. Only one of the hundred or so sperm that manage to make their way to the egg,

if it's there, will be able
to enter and fertilize it.
Which specific one, if any,
of the quarter of a billion
sperm that started that
journey will be able to prop-
agate itself, and whether it
will produce a male or a
female, was determined
long before the beginning
of carbon-based life on our
planet. As incredible as
that may seem, it is true
because everything as it is
at the moment the egg is
fertilized, was caused by
everything as it was the
moment before, and back
and back and back and back...

# Chance, and random, and spontaneous, and accidental.

Those are four words, some readers may argue, that show that *not everything* is *caused* and therefore *the one truth* is *not true*. But, contrary to what some may think, the first three words indicate actions without **_known, regularly occurring_** or **_apparent_** cause, not actions **_without_** cause. All actions that affect our world **_are caused_**.

The last word, accidental, means that something happened which wasn't **_planned_**; it doesn't mean something that wasn't **_caused_**. *All* accidents **_are caused_**.

**Speaking of *chance*,**
someone will tell you how
they miraculously bumped into
a friend in a foreign city clear
across the globe. They think
the odds of that naturally happening have to be so impossi-
bly high that its happening
has to be thought of as a *mir-
acle* and therefore, somehow,
*uncaused*. But the chance of
its happening is not at all
miraculous. If you search your
memories you will recall incidents like that in your own life.

I, myself, have had more than
a half dozen of these *miraculous* incidents. Once, some
years ago when I was in the
advertising business, I called in

a contractor (whose name I plucked out of the yellow pages) to come to my offices to break through a wall and install a door. He came on a Friday and with a crew of two helpers, installed the door in a matter of a few hours.

I discussed the project with him before he started, and, when he was done, paid him and thanked him. That night I flew to Florida with my family to spend a vacation at Disney World. The next morning while we were walking down Main Street at the Magic Kingdom a man suddenly stepped in front of me and with a look of disbelief said,

"Mr. MelGar! Remember me?" Well, I guess I don't have to tell you he was the contractor that I had seen for the first time the day before in New York —he was there with *his* family. After a few words he rushed back to them, got lost among the crowds and I never saw him again.

A mathematician could figure out the odds of something like that happening to all of us sometime in our lives. The odds would be long, but would be well short of miraculous; but *if* miraculous, and *caused* by a god, it would have been because of everything as it was at the beginning of everything.

# And, speaking of scientists, as I was earlier, here's a disclaimer:

*The one truth* is not an entirely new insight.

For thousands of years virtually all scientists and philosophers, whether professional or amateur, have known the *essentials* of *the one truth*. Mostly they referred and refer to it as *determinism,* or *causality,* or *causal determinism. The one truth*, however, is a more encompassing *form* of these, an *absolute* determinism, an *infinite* determinism that includes everything from the tiniest particle in nature to *the largest act of creation by a god* —a *form* those scientists and philosophers did not, or dared not, contemplate.

There are some *scientists* and *philosophers*, living today, who dismiss *any form* of determinism —especially those *forms* that claim to be absolute— and they profess to do so strictly on rational, reasoned, logical, scientific grounds.

Those *scientists* say that two of the three foundations of physics argue against the *absolute* truth of any form of determinism. They say, among other things, that there are aspects of *quantum mechanics* that argue against *everything* being caused, and that there are aspects of *Einstein's relativity* that argue against the statement that motion can only proceed *forward* in time.

*The philosophers*, on the other hand, appear, it seems to me, to oppose determinism because they don't feel determinism (with its concomitant assertion that we don't have *free* will) should be thought true because they believe it would be detrimental to our laws and morals to think it so, and they write intricately reasoned volumes asserting some scientifically deduced fault in it.

These technically complex, and difficult to grasp, negative arguments that have been advanced by a number of respected scientists and philosophers, are impossible to *present* and *respond* to equitably in this simple little book. Anyone who questions the truth that this book claims, and who wishes to know the opposing, *and supporting,* views, will find ample, serious literature under the headings of *determinism, causality, causal determinism,* or *free will.* (But be forewarned: be prepared for a search that will be arduous and difficult to navigate. How difficult will depend on the extent of your scientific knowledge and how much interest you have in the search and what pre-formed opinions you bring to it. It will require you to discern, to the best of your abilities, whether what is being advanced by the proposers are proven facts, hypotheses, or wishful beliefs —a difficult task when all may be proposed in the same context.

Be forewarned, also, that many in the *applied* sciences consider, *for the purposes of their work,* that something that is *not predictable* —something that is random or probabilistic or chaotic, for instance— *is not deterministic.* That does not mean that they don't acknowledge that everything as it is at this moment, predictable or not, was determined by everything as it was a moment ago.)

# And speaking of predictable:

*Nowhere in this book is it claimed that because everything was determined at the beginning, everything is **predictable**.*

That is not to say that short term, simple things are not predictable. For instance, you are standing out in the middle of the desert on a sunny, windless day; you have a cannon ball in your hand and you let it go. We can safely predict that the cannon ball will fall to the ground. We all know it will fall to the ground. We all know the law of physics that is called gravity; we know it because we experience it every time we fall.

And this is predictable: Get a pair of dice; hold them squared-up and level between two fingers; hold them a short ten centimeters above a smooth, hard, flat surface and drop them. The numbers that will end face up are the same numbers that were face up before you dropped the dice. That was predictable because it's simple: the distance the dice fell was too short to cause the dice to bounce or change position.

And while there are many other things that are simple and predictable, there are many more that, though causally determined, are much too complex to predict:

Get a large barrel and fill it to the brim with dice; take this barrel to the top of a high hill; dump all the dice down the hill's stone face; watch them roll down, tumbling, bouncing off each other and the hard rough surface of the hill; watch as they come to a halt at the flat ground at the bottom; look at the numbers that are face up (probably nearly equal amounts of each); they would all have been *determined* at the beginning, long before dice were thought of, but no gambler or mathematician could have predicted them.

Think of a maple tree in the autumn, its leaves all oranges and reds. Exactly when each of its individual leaves will fall to the ground —the random fall of each leaf dependent on a multitude of factors, including the actions of sun and rain and wind and bacteria, and the whims of squirrels and birds and bugs and bears— would be impossible to predict. Ask any arborist, or physicist, if they could predict *even one.*

But not everything that is unpredictable is so because of its complexity. In the sub-atomic world of quantum mechanics there is a simple scientific truth known as the *Heisenberg uncertainty principle*.

It states that we can't know *both* the *location* and the *momentum* of a sub-atomic particle at any given moment —the act of measuring its location, for example, would impact its momentum. And so, if we can't know both its location and momentum as it is *at a given moment,* we can't *predict* where it will be *a moment later* —though it will have been

## caused

by *where it was at that given moment* and by *how fast it was moving.*

# And one last example of the unpredictable:

The exact dialog that you, the reader, and I, the author, have had as you read this book —my words, your thoughts— was obviously determined before either of us was born; but neither I, nor anyone, could have predicted exactly what I would write, much less exactly what you have thought.

# And one final explanation:

"So," you may ask, since I haven't brought it up, "if we don't have free will and are therefore not responsible for what we do, how is it justified that we be punished if we commit a crime?"

Well, assuming we're not suffering from some mental deficiency, we *are responsible* for what we do, because none of us lives in a bubble. We all *know* what will happen to us if we get caught doing a crime because we learn from the punishment meted out to others who commit a crime. That's how we teach our children not to do bad things; we punish them when they do, so that they and their siblings will learn not to.

# *Epilogue:*

No one, no matter what they claim, *knows* whether everything "just is", or whether everything was created by a god who "just is". And no one, no matter what they claim, *knows* why everything exists rather than not exist (of course, if nothing existed, we wouldn't be here to know it). But there is one thing we *do know*: we know that everything as it is at this moment was caused by everything as it was a moment ago, and that's the one truth that governs everything; the only truth that governs man, the gods, the heavens and the universe as they unroll themselves forward to their inevitable, probability directed, causally determined, unpredictable future —whatever it may be.

www.ingramcontent.com/pod-product-compliance
Lightning Source LLC
Chambersburg PA
CBHW070028210526
45170CB00012B/306

\* 9 7 8 0 5 5 7 9 9 3 2 9 1 \*